Dinosaurs and Fossils

MATTHEW HUGO

TABLE OF CONTENTS

PIONEER VALLEY EDUCATIONAL PRESS, INC

DINOSAURS

Dinosaurs were animals that lived on the earth long, long ago.

Some dinosaurs were very small. They were no bigger than a chicken.

Some dinosaurs were very large.
They were even bigger
than an elephant.

TERRIBLE LIZARDS

The word *dinosaur* means "terrible lizard."

Dinosaurs were **reptiles**. Baby dinosaurs **hatched** from eggs just like snakes and alligators and lizards.

TYPES OF DINOSAURS

Some dinosaurs walked on two legs.
Some dinosaurs walked on four legs.
Some dinosaurs flew in the air
and some swam in the water.

Dinosaurs that walked on two legs

deinocheirus

tyrannosaurus rex

Dinosaurs that walked on four legs

triceratops

kentrosaurus

Most dinosaurs were plant eaters.
Many plant-eating dinosaurs
walked on four legs.
Most had small heads and long tails.
Many had **spikes** on their backs.

Triceratops was a plant-eating dinosau

Some dinosaurs were meat eaters.
Many meat-eating dinosaurs
walked on two legs.
They had sharp teeth.
They had short arms
with claws that helped them
grab their **prey**.

Tyrannosaurus rex was a meat-eating dinosaur.

WHAT HAPPENED TO THE DINOSAURS?

No one knows
what happened to the dinosaurs.

Some people think the dinosaurs
died from a big volcano.

Some people think the dinosaurs died when a large rock fell to Earth from space.

Some people think the dinosaurs died when the weather got too cold.

DINOSAUR FOSSILS

We can learn about dinosaurs by looking at dinosaur **fossils**. Fossils are parts of a plant or animal that have turned to stone after many, many years.

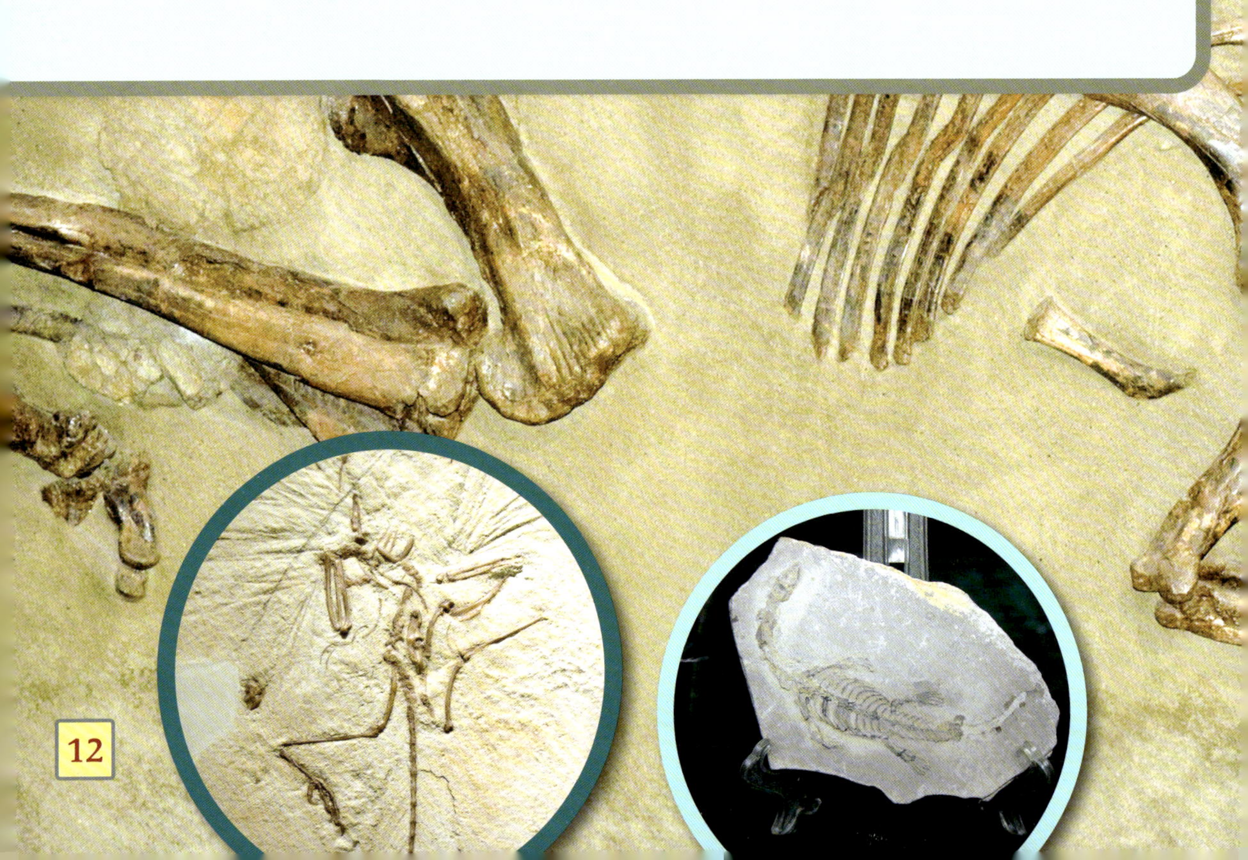

How Fossils Are Formed

1. A dinosaur dies.
2. Its body is buried with dirt, leaves, or sand.
3. The dinosaur bones turn into stone.
4. The bones break down, leaving a shape where they were.
5. The shape fills with minerals.
6. Millions of years later, we find the fossils on Earth's surface.

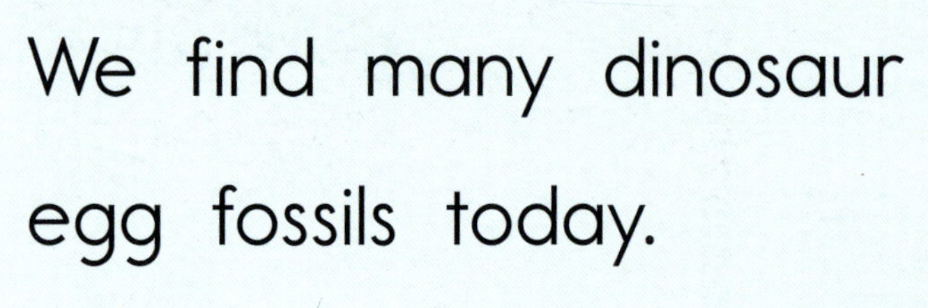

We find many dinosaur
egg fossils today.
Dinosaurs covered their eggs
with leaves or sand
to keep them warm.
Some of the eggs never hatched
and became buried.
Today they are fossils.

SAURS
CRETACEOUS
145 million years ago
Maiasaurus
Velociraptor
Allosaurus
Tyrannosaurus Rex
Troodon
Ankylosaurus
Triceratops
urus

DINOS

TRIASSIC
230 million years ago
JURASSIC
200 million years ago
Archaeopteryx
Coelophysis
Diplodocus
Plateosaurus
Brachiosaurus
Herrerasaurus
Heterodontosaurus
Stegos
Eoraptors

GLOSSARY

hatch
to come out of an egg

prey
an animal that is hunted for food by another animal

reptile
an animal that is cold-blooded and lays eggs like a lizard, alligator, or snake

spikes
sharp points